# Workbook

# MyMaths
## for Key Stage 3

powered by MyMaths.co.uk

OXFORD
UNIVERSITY PRESS

**UNIVERSITY PRESS**

Great Clarendon Street, Oxford, OX2 6DP, United Kingdom

Oxford University Press is a department of the University of Oxford.
It furthers the University's objective of excellence in research,
scholarship, and education by publishing worldwide. Oxford is a
registered trade mark of Oxford University Press in the UK and in
certain other countries

© Oxford University Press 2014

The moral rights of the authors have been asserted

First published in 2014

British Library Cataloguing in Publication Data
Data available

978-0-19-830471-5

14

Paper used in the production of this book is a natural, recyclable
product made from wood grown in sustainable forests.
The manufacturing process conforms to the environmental
regulations of the country of origin.

Printed in Great Britain by Ashford Print and Publishing Services,
Gosport

**1**  What does the circled digit stand for in each number?

   **a**  1⑧5            8 tens or eighty

   **b**  ⑤7                        **c**  21⑤

   **d**  ①892                      **e**  1⑤20

**2**  Here are three digits:

   **a**  What is the smallest number you can make?

   **b**  What is the largest number you can make?

**3**  What numbers are shown here?

   **a**  | 100 | 10 | 10 | 1 | 1 | 1 |    _____

   **b**  | 100 | 100 | 100 | 10 | 10 | 1 | 1 |    _____

   **c**  | 100 | 100 | 100 | 10 | 1 | 1 | 1 |    _____

**4**  Write these numbers as 100s, 10s and units on the cards like question 3.

   **a**  142

   **b**  231

   **c**  420

   **d**  205

# 1e Adding decimals

 I can do this page!

**1** Write each amount as a decimal.

| | | | | | | |
|---|---|---|---|---|---|---|
| (10 pence) | (20 pence) | (one penny) | (five pence) | (two pence) | (£1) | (fifty pence) |
| £0.10 | £0.20 | | | | | |

**2** Draw these amounts in the boxes using exactly five coins.

**a** £1.25

**b** £1.52

**c** £1.34

**3** Add these amounts.

**a** £0.10 + £0.50 = _____  **b** £0.10 + £0.80 = _____

**c** £0.10 + £0.30 = _____  **d** £0.10 + £0.40 = _____

**e** £0.10 + £0.70 = _____  **f** £0.10 + £0.20 = _____

**g** £0.20 + £0.40 = _____  **h** £0.40 + £0.50 = _____

**i** £0.50 + £0.20 = _____  **j** £0.30 + £0.50 = _____

**k** £0.40 + £0.20 = _____  **l** £0.70 + £0.20 = _____

MyMaths.co.uk

Q 1226, 1377  SEARCH

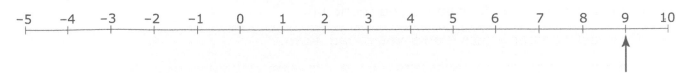

I can do this page!

1 On this number line the arrow points to position 9.

-5  -4  -3  -2  -1  0  1  2  3  4  5  6  7  8  9  10

Draw arrows to point to these positions.
Label the arrows with their letters.

a 5        b −3        c −1        d 8        e 0

2 Thermometers measure temperature.

Write the missing temperatures on these thermometers.

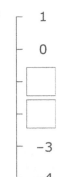
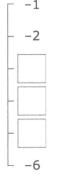

3 Circle the lower temperature in each pair.

a −3°C    4°C              b 2°C    4°C

c 6°C    0°C               d 3°C    −1°C

e −1°C    1°C              f 5°C    −5°C

4 Complete this number line that starts at -10.

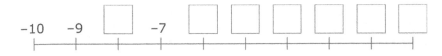

-10  -9  [ ]  -7  [ ] [ ] [ ] [ ] [ ] [ ]

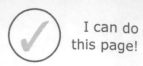

When rounding to the nearest 10

- if the last number is 5 or above you round up

- if the last number is less then 5 you round down.

| 71  72  73  74 | 75  76  77  78  79 |
|---|---|
| These will round down to 70. | These will round up to 80. |

**1** Circle the numbers that would round down.

23   98   73   81   37   59   65   42   94   26   46   51

**2** Circle the numbers that would round up.

53   78   26   55   94   24   18   83   16   82   49   37

**3** Round these numbers down.

**a** 72 ⟶ ☐   **b** 51 ⟶ ☐   **c** 34 ⟶ ☐   **d** 21 ⟶ ☐

**e** 13 ⟶ ☐   **f** 82 ⟶ ☐   **g** 63 ⟶ ☐   **h** 44 ⟶ ☐

**4** Round these numbers up.

**a** 79 ⟶ ☐   **b** 57 ⟶ ☐   **c** 36 ⟶ ☐   **d** 85 ⟶ ☐

**e** 68 ⟶ ☐   **f** 25 ⟶ ☐   **g** 17 ⟶ ☐   **h** 48 ⟶ ☐

**5** Round these amounts to the nearest £10.

**a** £88 ⟶ ☐   **b** £51 ⟶ ☐   **c** £14 ⟶ ☐

**d** £75 ⟶ ☐   **e** £12 ⟶ ☐   **f** £47 ⟶ ☐

**g** £33 ⟶ ☐   **h** £76 ⟶ ☐   **i** £29 ⟶ ☐

**j** £64 ⟶ ☐   **k** £48 ⟶ ☐   **l** £22 ⟶ ☐

MyMaths.co.uk

1   Write these units of measurement in the correct list.

| month | kilogram | second | centimetre | mile |
| year | minute | tonne | week | kilometre |
| hour | gram | millimetre | day | metre |

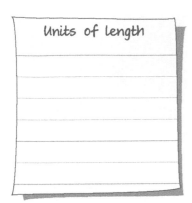
Units of length

Units of time

Units of weight

2   Measure these lines using a ruler.

a _____     _____ cm

b _____     _____ cm

c _____     _____ cm

d _____     _____ cm

e _____     _____ cm

3   Measure the length and width of this rectangle. Write your answers on the diagram.

_____ cm

_____ cm

1 Draw lines of these lengths above the rulers.

a 3.4 cm

| 0 cm | 1 | 2 | 3 | 4 | 5 | 6 | 7 | 8 | 9 | 10 |

b 6.7 cm

| 0 cm | 1 | 2 | 3 | 4 | 5 | 6 | 7 | 8 | 9 | 10 |

c 9.1 cm

| 0 cm | 1 | 2 | 3 | 4 | 5 | 6 | 7 | 8 | 9 | 10 |

2 What reading does each scale show?
Write your answer as a decimal.

A = _____

B = _____

C = _____

D = _____

E = _____

F = _____

G = _____

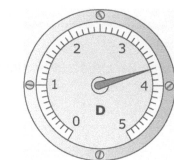

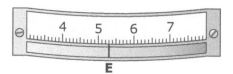

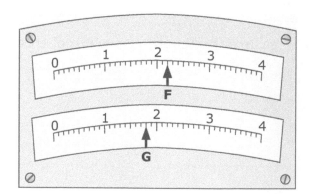

MyMaths.co.uk

Q 1232  SEARCH

## 2c Time

**1** Put hands on each clock to show the time.

| 9 o'clock | half past one | 7:15 | quarter to five | five past six |
|---|---|---|---|---|
|  |  |  |  |  |

**2** What times are shown on these clocks?

_____    _____    _____    _____    _____

**3** What times are shown on these clocks? Use **a.m.** for times before midday and **p.m.** for times after midday.

_____    _____    _____

_____    _____

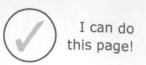

I can do
this page!

1   What is the perimeter of each shape?

**a**

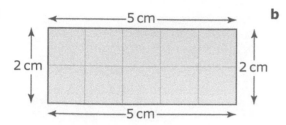

Perimeter = _____ cm

**b**

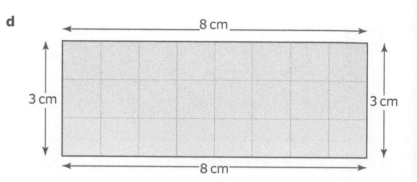

Perimeter = _____ cm

**c**

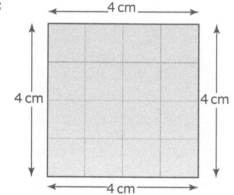

Perimeter = _____ cm

**d**

Perimeter = _____ cm

2   Measure the perimeter of each shape with a ruler.

**a**

Perimeter = _____ cm

**b**

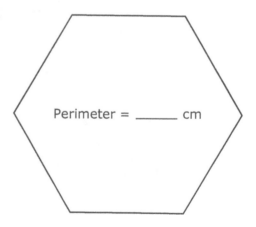

Perimeter = _____ cm

**c**

Perimeter = _____ cm

**d**

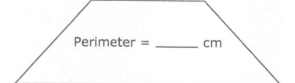

Perimeter = _____ cm

**MyMaths**.co.uk      🔍 1110      SEARCH

This is a plan of the ground floor of a palace.

The measurements of each room are in metres.

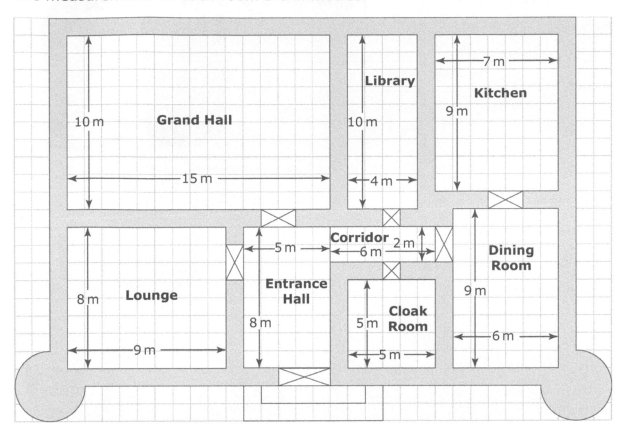

1   What is the perimeter of each room?

   **a**  The Cloak Room  \_\_\_\_\_ m          **b**  The Library       \_\_\_\_\_ m

   **c**  The Lounge      \_\_\_\_\_ m          **d**  The Kitchen      \_\_\_\_\_ m

   **e**  The Dining Room  \_\_\_\_\_ m          **f**  The Grand Hall    \_\_\_\_\_ m

2   What is the area of each room? Count the squares.

   **a**  The Cloak Room  \_\_\_\_\_ m$^2$        **b**  The Library      \_\_\_\_\_ m$^2$

   **c**  The Lounge      \_\_\_\_\_ m$^2$        **d**  The Kitchen     \_\_\_\_\_ m$^2$

   **e**  The Dining Room  \_\_\_\_\_ m$^2$        **f**  The Grand Hall   \_\_\_\_\_ m$^2$

   **g**  The Entrance Hall \_\_\_\_\_ m$^2$        **h**  The Corridor    \_\_\_\_\_ m$^2$

1   The soda can contains 70 ml.

The cola can contains 75 ml.

a   Which can contains more? _____

b   How much more? _____

2   The butter weighs 200 g.

The margarine weighs 250 g.

a   Which weighs more? _____

b   How much more? _____

3   a   Which roll of tape is the longest? _____

b   Which roll of tape is the shortest? _____

c   How much longer is B than C? _____

d   How much longer is A than B? _____

e   How much shorter is C than A? _____

A = 33 m     B = 20 m

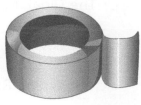

C = 12 m

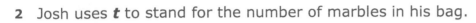

**1**  Josh keeps some of his marbles in a bag.

Can you tell exactly how many marbles are in Josh's bag, without guessing?

Circle the correct answer:   Yes   or   No

**2**  Josh uses **t** to stand for the number of marbles in his bag.

He has **t** marbles in the bag and another 4 marbles.

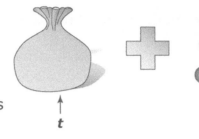

$t$

$4$

**a**  How many marbles does Josh have altogether?

Circle the correct answer.

$t - 4$ marbles          $t + 4$ marbles          $4 - t$ marbles

**b**  Josh won 6 more marbles.

How many marbles does he now have altogether?

Circle the correct answer.

$t + 4$ marbles          10 marbles          $t + 10$ marbles          2 marbles

**3  a**  These parcels weigh **n** kilograms altogether.

The small parcel to the side weighs 8 kilograms.

Add the small parcel to the big pile.

What is the total weight in kg?

Circle the correct answer.

$n + 8$     $n - 8$     $8 - n$     $n \times 8$

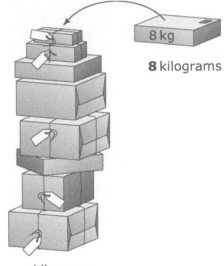

8 kg

**8** kilograms

**n** kilograms

**b**  Add another parcel weighing 12 kilograms to the pile.

What is the new total weight in kg?

Circle the correct answer.

$n + 12$     12     20     $n + 20$

1   Ravi has **n** apples.

Toby has 7 more apples than Ravi.

How many does Toby have?

Toby has **n** + 7 apples.

Jenny has 3 more apples than Ravi.

How many does Jenny have?

Jenny has _____ apples.

2   Tom, Jade and Eva all had the same number
of sweets.

Tom ate 5 sweets so he had **n** − 5 sweets left.

Jade ate 3 sweets so she had _____ sweets left.

Eva ate 6 sweets so she had _____ sweets left.

3   The Jones family had a chocolate bar with **n** pieces.

Dad ate 3 pieces leaving **n** − 3.

Mum ate 2 pieces leaving **n** − 3 − 2 or **n** − 5.

Clare ate 5 pieces leaving _____ or _____

Paul ate 3 pieces leaving _____ or _____.

MyMaths.co.uk          🔍 1158     SEARCH

## Number towers

Add the numbers in two boxes next to each other to get the number in the box below.

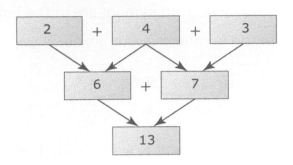

Complete these towers.

**1**

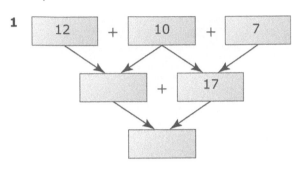

**2**

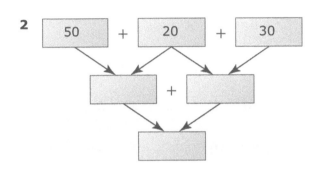

**3**

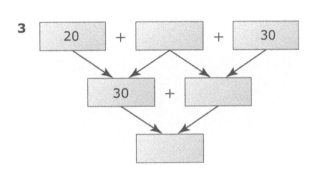

**4**

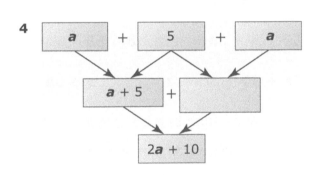

**5**

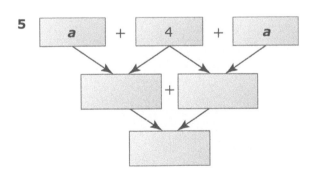

**6**

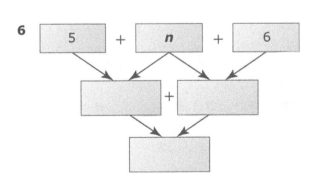

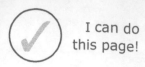 I can do this page!

1   How many of each symbol are in the box?

Write the numbers in the boxes below.

$\boxed{\phantom{0}}\,a + \boxed{\phantom{0}}\,g + \boxed{\phantom{0}}\,k$

| a | k | g | a | a |
|---|---|---|---|---|
|   | k | k | g | g |
| a | a | a | g | a |
|   | k | a | g | a | g |

2   How many **a** symbols are in each line?

Write the number in the space.

The first one is done for you

a   $b + a + 2a = \underline{\ \ 3\ \ } \ a + b$

b   $4a + 2a - b = \underline{\ \ \ \ \ } \ a - b$

c   $6a - a + 3b = \underline{\ \ \ \ \ } \ a + 3b$

d   $2a + 2a + 2b = \underline{\ \ \ \ \ } \ a + 2b$

e   $5b + a + 3a = \underline{\ \ \ \ \ } \ a + 5b$

f   $2a + 3b + 3b - a = \underline{\ \ \ \ \ } \ a + 6b$

3   How many of each symbol are in each line?

Write the numbers in the spaces.

The first one is done for you

a   $5p + 4r - r = \underline{\ \ 5\ \ } \ p + \underline{\ \ 3\ \ } \ r$

b   $8n + 3n + 6t = \underline{\ \ \ \ \ } \ n + \underline{\ \ \ \ \ } \ t$

c   $2a + a + 3b + b = \underline{\ \ \ \ \ } \ a + \underline{\ \ \ \ \ } \ b$

d   $2g + 3s + 2g + 3s = \underline{\ \ \ \ \ } \ g + \underline{\ \ \ \ \ } \ s$

e   $6c - c + 3d - d = \underline{\ \ \ \ \ } \ c + \underline{\ \ \ \ \ } \ d$

# 3e Substitution

I can do this page!

Small boxes hold 3 doughnuts.

Medium boxes hold 6 doughnuts.

Large boxes hold 10 doughnuts.

You can use symbols to represent the numbers of doughnuts in each box:

$s = 3$ $\qquad\qquad\qquad$ $m = 6$ $\qquad\qquad\qquad$ $l = 10$

For a small box and two extra doughnuts:

$s + 2$ $\qquad$ means $\qquad$ $3 + 2$

so $\quad s + 2 = 5$

**1** Draw $s + 1$ doughnuts:

**2** Draw $s - 2$ doughnuts:

**3** Draw $m + 3$ doughnuts:

**4** Draw $m - 1$ doughnuts:

**5** Draw $m - 3$ doughnuts:

**6** Draw $l - 3$ doughnuts:

**7** Draw $s + s$ doughnuts:

**8** Draw $s + m$ doughnuts:

A dairy farm is a business. The farmer sells the cows' milk to make a profit, so they must be well looked after.

**Task 1**

Put the cows in order by weight, lightest first.

Perfect for drinking

1st Class Butter

The best cheese

**HOLSTEIN**
Weight: 700 kg
Food: 24 kg per day
Number on farm: 25

**JERSEY**
Weight: 500 kg
Food: 20 kg per day
Number on farm: 10

**BROWN SWISS**
Weight: 600 kg
Food: 25 kg per day
Number on farm: 15

**Task 4**

How much does the food cost per day

a in summer

b in winter?

**Task 2**

How many cows are there in total in the herd?

**Task 3**

How much does the herd eat per day?

*** **FARM FOOD** ***

Grass ... Free (April-Sept only)

Silage ... £10 per 100 kg

PLEASE RETAIN RECEIPT
THANK YOU.

# 4a Writing fractions

**1** Complete these sentences by looking at each shape. The first one is done for you.

**a**

There are **3** parts.

1 part is shaded.

$\frac{1}{3}$ is shaded.

**b**
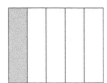

There are _____ parts.

_____ part is shaded.

_____ is shaded.

**c**
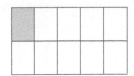

There are _____ parts.

_____ part is shaded.

_____ is shaded.

**d**

There are _____ parts.

_____ part is shaded.

_____ is shaded.

**e**

There are _____ parts.

_____ part is shaded.

_____ is shaded.

**f**
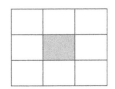

There are _____ parts.

_____ part is shaded.

_____ is shaded.

**2**

This shape is divided into 3 parts.

2 parts are shaded.

$\frac{2}{3}$ of the whole shape is shaded.

What fraction of each shape is shaded?

**a**
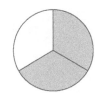
_____

**b**

_____

**c**
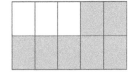
_____

**d**
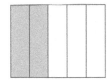
_____

**e**
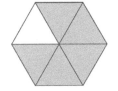
_____

**f**
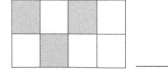
_____

**1** Shade $\frac{1}{2}$ of each shape.

Write the fraction that is equivalent to $\frac{1}{2}$ in each drawing.

**a**

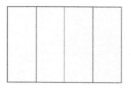

$\frac{1}{2}$ is the same as ☐

**b**

$\frac{1}{2}$ is the same as ☐

**c**

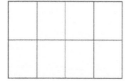

$\frac{1}{2}$ is the same as ☐

**d**

$\frac{1}{2}$ is the same as ☐

**2** Shade $\frac{1}{5}$ of this shape.

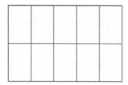

$\frac{1}{5}$ is the same as ☐

**3** Shade $\frac{1}{4}$ of this shape.

$\frac{1}{4}$ is the same as ☐

**4** Shade $\frac{1}{6}$ of this shape.

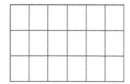

$\frac{1}{6}$ is the same as ☐

**5** Shade $\frac{1}{5}$ of this shape.

$\frac{1}{5}$ is the same as ☐

**MyMaths**.co.uk  🔍 1371  SEARCH

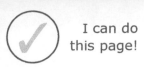
1   Here are 10 cubes. Put them in two equal groups.

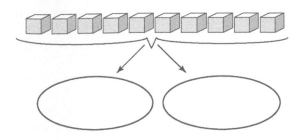

What is $\frac{1}{2}$ of 10?

Answer: _____

2   Here are 15 counters. Put them in three equal groups.

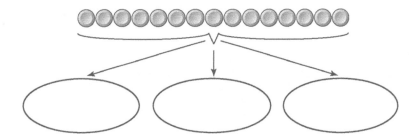

What is $\frac{1}{3}$ of 15?

Answer: _____

3   Which division goes with each calculation?

Draw arrows to connect them.

Calculation                    division

$\frac{1}{4}$ of _____            ÷ 2

$\frac{1}{3}$ of _____            ÷ 6

$\frac{1}{8}$ of _____            ÷ 4

$\frac{1}{6}$ of _____            ÷ 5

$\frac{1}{2}$ of _____            ÷ 10

$\frac{1}{10}$ of _____           ÷ 8

$\frac{1}{5}$ of _____            ÷ 3

For example:
$\frac{1}{4}$ of 12 means 12 ÷ 4 = 3

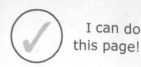
**1** Use the multiplication grid to help you find:

**a** $\frac{1}{2}$ of 8  = _____

**b** $\frac{1}{3}$ of 9  = _____

**c** $\frac{1}{4}$ of 16 = _____

**d** $\frac{1}{5}$ of 15 = _____

**e** $\frac{1}{3}$ of 18 = _____

**f** $\frac{1}{4}$ of 24 = _____

**g** $\frac{1}{6}$ of 24 = _____

**h** $\frac{1}{2}$ of 12 = _____

**i** $\frac{1}{5}$ of 30 = _____

**j** $\frac{1}{6}$ of 30 = _____

| ×  | 1 | 2  | 3  | 4  | 5  | 6  |
|----|---|----|----|----|----|----|
| 1  | 1 | 2  | 3  | 4  | 5  | 6  |
| 2  | 2 | 4  | 6  | 8  | 10 | 12 |
| 3  | 3 | 6  | 9  | 12 | 15 | 18 |
| 4  | 4 | 8  | 12 | 16 | 20 | 24 |
| 5  | 5 | 10 | 15 | 20 | 25 | 30 |
| 6  | 6 | 12 | 18 | 24 | 30 | 36 |

**2** Find:

**a** $\frac{1}{2}$ of 18 = _____

**b** $\frac{1}{2}$ of 24 = _____

**c** $\frac{1}{2}$ of 40 = _____

**3** Find:

**a** $\frac{1}{4}$ of 28 = _____

**b** $\frac{1}{4}$ of 40 = _____

**c** $\frac{1}{4}$ of 32 = _____

**4** Find:

**a** $\frac{1}{3}$ of 24 = _____

**b** $\frac{1}{3}$ of 30 = _____

**c** $\frac{1}{3}$ of 36 = _____

**5** Use the machines to find these fractions of amounts.

**a** $\frac{3}{4}$ of 12

12 → ÷ 4 → ☐ → × 3 → = ☐

**b** $\frac{3}{4}$ of 20

20 → ÷ 4 → ☐ → × 3 → = ☐

**c** $\frac{2}{3}$ of 15

15 → ÷ 3 → ☐ → × 2 → = ☐

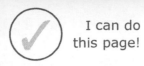
1   Here are 20 counters:

Share them equally into the percentage strip.

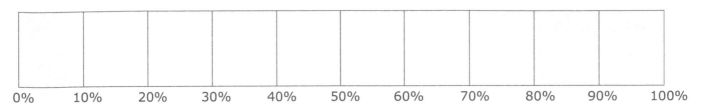

0%   10%   20%   30%   40%   50%   60%   70%   80%   90%   100%

Use the strip to work out:

a   50% of 20 counters = _____      b   10% of 20 counters = _____

c   30% of 20 counters = _____      d   80% of 20 counters = _____

2   A pizza is cut into 10 equal slices. Each slice is 10% of the whole pizza.

a   Colour 3 slices blue. What percentage is this?     _____%

b   Colour 3 slices green. What percentage is this?    _____%

c   Colour 3 slices red. What percentage is this?      _____%

d   What percentage of the pizza is left?              _____%

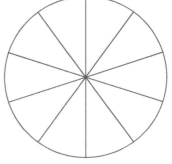

3   Here are 50 counters:

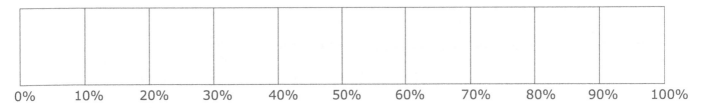

Share them equally into the percentage strip.

0%   10%   20%   30%   40%   50%   60%   70%   80%   90%   100%

Use the strip to work out:

a   10% of 50 counters = _____      b   30% of 50 counters = _____

c   50% of 50 counters = _____      d   80% of 50 counters = _____

**1** Here are 20 beads.

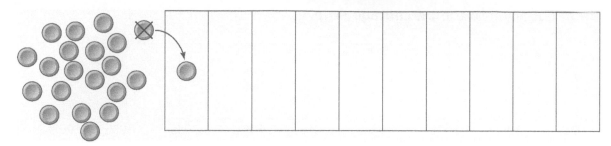

**a** Share the beads equally between the 10 sections.

Each section is 10% or $\frac{1}{10}$.

**b** Use the drawing to link these calculations to their correct answers. The first one is done for you.

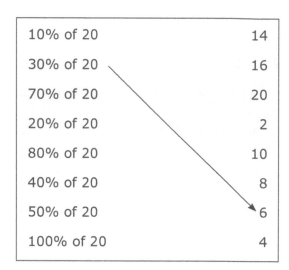

| | |
|---|---|
| 10% of 20 | 14 |
| 30% of 20 | 16 |
| 70% of 20 | 20 |
| 20% of 20 | 2 |
| 80% of 20 | 10 |
| 40% of 20 | 8 |
| 50% of 20 | 6 |
| 100% of 20 | 4 |

**2** This plank measures 10 cm.

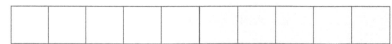

10%

**a** Divide the plank into exactly 10% sections. The first one has been done for you.

**b** 10% of the plank has been shaded. Shade another 40%.

**c** How much of the plank is now shaded? Answer: _____ %

**3 a** Shade 60% of this rectangle.

**b** Use the diagram in question 1 to find 60% of 20.   Answer: _____

**MyMaths**.co.uk    Q 1030    SEARCH

I can do this page!

1 How much of each shape is shaded?

Give your answer as a fraction, decimal and percentage. The first is done for you.

a

$\frac{7}{10}$ or 0.7 or 70%

b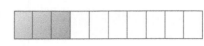

_____ or _____ or _____

c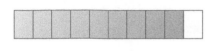

_____ or _____ or _____

d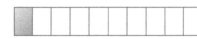

_____ or _____ or _____

e

_____ or _____ or _____

f

_____ or _____ or _____

2 Shade in these parts of this 10 × 10 square.

The first has been done for you.

A $\frac{4}{100}$ or 0.04 or 4%     B $\frac{7}{100}$ or 0.07 or 7%

C $\frac{5}{100}$ or 0.05 or 5%     D $\frac{10}{100}$ or 0.1 or 10%

E $\frac{14}{100}$ or 0.14 or 14%     F $\frac{17}{100}$ or 0.17 or 17%

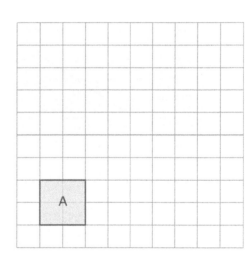

3 Write these fractions as decimals and percentages.

The first one is done for you.

a $\frac{6}{100}$ = 0.06 = 6%     b $\frac{10}{100}$ = _____ = _____     c $\frac{16}{100}$ = _____ = _____

d $\frac{25}{100}$ = _____ = _____     e $\frac{65}{100}$ = _____ = _____     f $\frac{40}{100}$ = _____ = _____

4 Write these percentages as decimals and fractions.

The first one is done for you.

a 2% = 0.02 = $\frac{2}{100}$     b 15% = _____ = _____     c 35% = _____ = _____

d 80% = _____ = _____     e 33% = _____ = _____     f 90% = _____ = _____

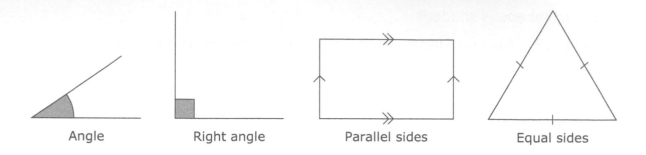

Angle     Right angle     Parallel sides     Equal sides

**1** Mark the angles and parallel or equal sides in these shapes.

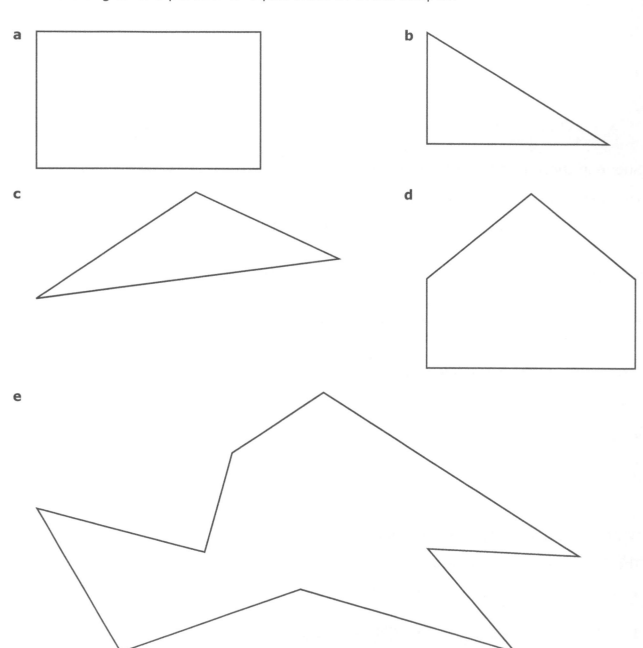

a

b

c

d

e

MyMaths.co.uk     🔍 1082    SEARCH

I can do
this page!

**1** What is the size of each angle?

Give your answer in degrees (°).

**a**

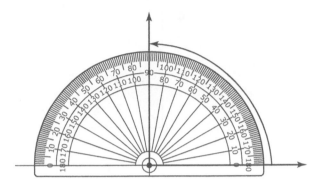

**b**

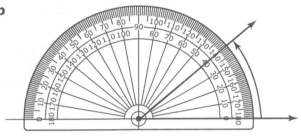

**c**

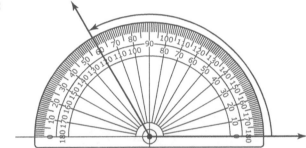

**d**

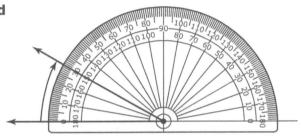

**e**

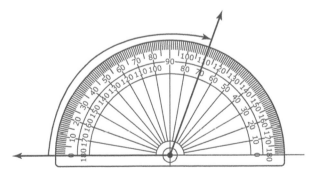

**f**

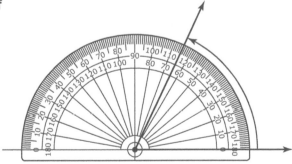

1   There are 90° in a right angle.

Calculate the size of the missing angle in each right-angle.

**a**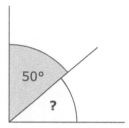

$$50 + ? = 90°$$

$$? = \underline{\hspace{2cm}}$$

**b**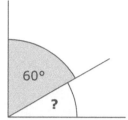

$$60° + ? = \underline{\hspace{2cm}}$$

$$? = \underline{\hspace{2cm}}$$

**c**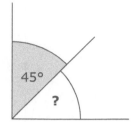

$$45° + ? = \underline{\hspace{2cm}}$$

$$? = \underline{\hspace{2cm}}$$

2   There are 180° on a straight line.

Calculate the size of the missing angle.

**a**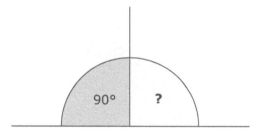

$$90° + ? = 180°$$

$$? = \underline{\hspace{2cm}}$$

**b**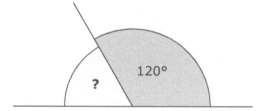

$$120° + ? = \underline{\hspace{2cm}}$$

$$? = \underline{\hspace{2cm}}$$

**c**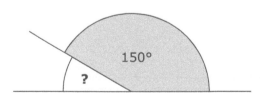

$$150° + ? = \underline{\hspace{2cm}}$$

$$? = \underline{\hspace{2cm}}$$

MyMaths.co.uk    Q 1082    SEARCH

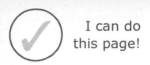

I can do
this page!

Franco goes on a journey. Here is a map of the area.

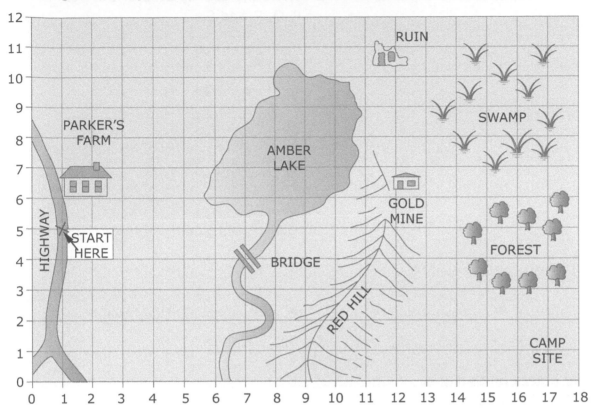

**1** Plot the points and join them on the map to show Franco's journey.

(1,5)  (3,7)  (8,11)  (11,10)  (11,8)  (13,7)

(14,5)  (14,3)  (15,2)  (17,1)

> Remember, you go **across** first, then up.

**2** What will Franco see on his left at (3,7)?  _____

**3** What will Franco see on his right at (11,10)?  _____

**4** What will Franco see on his left at (11,10)?  _____

**5** What will Franco see on his right at (13,7)?  _____

**6** What will Franco see on his left at (15,2)?  _____

**7** Plot a shorter route from the Highway (1,5) to the Camp Site (17,1).
Draw the new route on the map, and write the coordinates here.

_____

_____

**1** **a** Write the coordinates of the points A, B, C and D.

**b** Write the coordinates of the points P, Q, R and S.

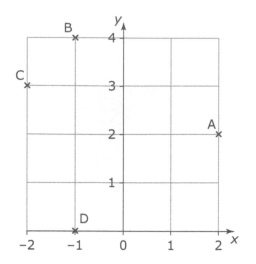

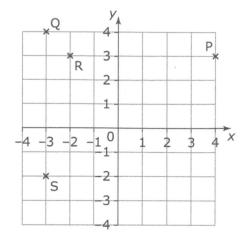

A(_____ , _____)   B(_____ , _____)

C(_____ , _____)   D(_____ , _____)

P(_____ , _____)   Q(_____ , _____)

R(_____ , _____)   S(_____ , _____)

**2** **a** Plot these points on this grid.
(2, −2)   (−1, 2)   (−5, −1)   (−2, −5)

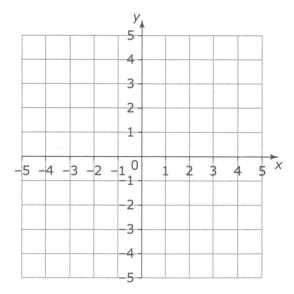

**b** Join all the points together, in order, with straight lines.

**c** What shape have you drawn? _____

**MyMaths**.co.uk   Q 1093   SEARCH

## 6c  Reading graphs

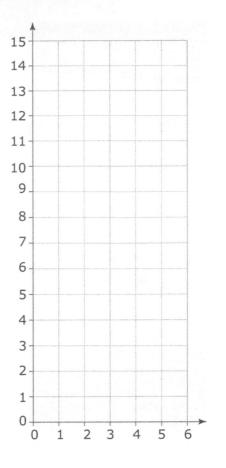

**I can do this page!**

1  **a**  Write coordinate pairs from this mapping.

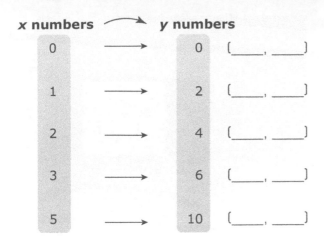

| x numbers | | y numbers | |
|---|---|---|---|
| 0 | → | 0 | (____, ____) |
| 1 | → | 2 | (____, ____) |
| 2 | → | 4 | (____, ____) |
| 3 | → | 6 | (____, ____) |
| 5 | → | 10 | (____, ____) |

**b**  Plot the points on this grid.

**c**  Join the points with a straight line. Use a ruler.

**d**  Write the rule that connects the **before** and **after** numbers.

before ──→ ▶ × ____ ──→ after

**e**  Fill in these coordinate pairs from the graph:

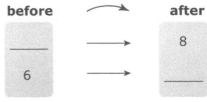

| before | → | after |
|---|---|---|
| ____ | → | 8 |
| 6 | → | ____ |

2  **a**  Complete the mapping by filling in the **after** values from the graph.

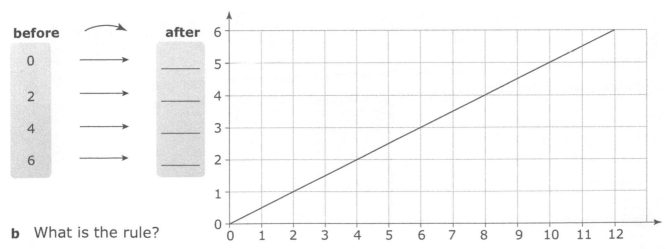

| before | | after |
|---|---|---|
| 0 | → | ____ |
| 2 | → | ____ |
| 4 | → | ____ |
| 6 | → | ____ |

**b**  What is the rule?

_____

**MyMaths**.co.uk

29

# Case Study 2: Recycling and energy

Each aluminium drink can that is recycled saves enough energy to run a television set for three hours.

**3 hours**

## Task 1

**a** How many hours of television would you power by recycling:

  i   2 aluminium cans

  ii  3 aluminium cans

  iii 4 aluminium cans

  iv  10 aluminium cans?

**b** How many cans would save enough energy for:

  i   6 hours of TV

  ii  12 hours of TV

  iii 15 hours of TV

  iv  24 hours of TV?

Making 1 new aluminium can uses the same amount of energy as recycling 20 cans.

## Task 2

How many recycled cans could you make using the same amount of energy as:

**a** 2 new cans

**b** 5 new cans?

Roughly how many cans do you drink in a week? How many hours of television could you power by recycling those cans?

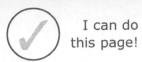

Use these number lines to help you add these numbers.
The first one is done for you.

**1**  13 + 14 = 27

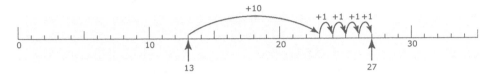

**2**  20 + 8 = _____

**3**  50 + 18 = _____

**4**  40 + 29 = _____

**5**  23 + 20 = _____

**6**  10 + 19 = _____

**7**  40 + 32 = _____

**8**  26 + 33 = _____

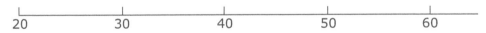

**9**  11 + 29 = _____

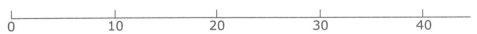

1  Circle the largest number in each box.
   Find the difference between the numbers.

| a | | |
|---|---|---|
| | 8 | (12) |

Difference = __4__

| b | | |
|---|---|---|
| | 9 | 15 |

Difference = _____

| c | | |
|---|---|---|
| | 11 | 7 |

Difference = _____

| d | | |
|---|---|---|
| | 13 | 9 |

Difference = _____

| e | | |
|---|---|---|
| | 14 | 7 |

Difference = _____

| f | | |
|---|---|---|
| | 7 | 13 |

Difference = _____

| g | | |
|---|---|---|
| | 9 | 15 |

Difference = _____

| h | | |
|---|---|---|
| | 16 | 8 |

Difference = _____

| i | | |
|---|---|---|
| | 4 | 10 |

Difference = _____

| j | | |
|---|---|---|
| | 13 | 8 |

Difference = _____

| k | | |
|---|---|---|
| | 16 | 7 |

Difference = _____

| l | | |
|---|---|---|
| | 6 | 11 |

Difference = _____

2  Subtract these numbers.

a  $14 - 10 =$ _____    b  $16 - 10 =$ _____    c  $13 - 10 =$ _____

d  $17 - 10 =$ _____    e  $11 - 10 =$ _____    f  $15 - 10 =$ _____

g  $12 - 10 =$ _____    h  $18 - 10 =$ _____    i  $15 - 5 =$ _____

j  $13 - 3 =$ _____    k  $17 - 7 =$ _____    l  $12 - 2 =$ _____

m  $19 - 9 =$ _____    n  $14 - 4 =$ _____    o  $11 - 1 =$ _____

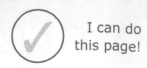

MyMaths.co.uk

Q 1055, 1224  SEARCH

# 8c  Reading lists and tables

I can do this page!

1  Sort these 20 words into the four lists.

| hammer | France | Emily | blue | Ann | Canada |
| hack-saw | green | brown | James | China | yellow | John |
| drill | Egypt | purple | Alfie | spade | Pakistan | pliers |

| **Colours** | **Names** | **Tools** | **Countries** |
|---|---|---|---|
| _____ | _____ | _____ | _____ |
| _____ | _____ | _____ | _____ |
| _____ | _____ | _____ | _____ |
| _____ | _____ | _____ | _____ |
| _____ | _____ | | |

2  This is Joseph's receipt from the Supermarket.

a  How many tins did Joseph buy? _____

b  How much did he spend on vegetables? _____

c  How much did he spend on meat? _____

d  How much did he spend on tins? _____

e  On which kind of product did he spend the **least**? _____

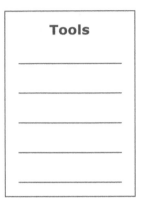

| Super Supermarket | |
|---|---|
| Vegetables | £1.25 |
| Vegetables | £0.50 |
| Vegetables | £0.75 |
| Meat | £3.50 |
| Meat | £2.50 |
| Tin | £1.20 |
| Tin | £1.20 |
| Tin | £1.20 |
| Total | £____ |

**MyMaths**.co.uk

33

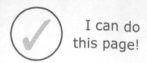

Class 7P call out their favourite colours for a survey.

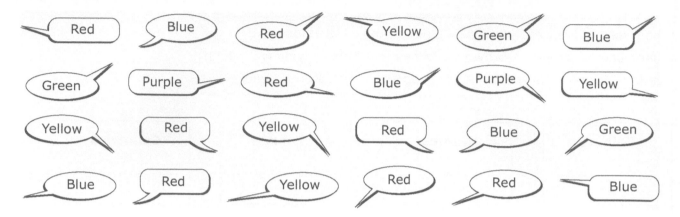

1  Count how many people called out each colour and write the numbers in the table. The first one has been done for you.

| Colour | Number of people |
|--------|------------------|
| Blue   | 6                |
| Green  |                  |
| Purple |                  |
| Red    |                  |
| Yellow |                  |

2  Complete this pictogram to show this data.
   Use ☺ to represent one person.

3  How many students called out Blue?

   _____

4  What was the **most** popular colour?

   _____

5  What was the **least** popular colour?

   _____

**Favourite colour**

| | |
|--------|---|
| Blue   | |
| Green  | |
| Purple | |
| Red    | |
| Yellow | |
| | Total number of students |

Key: ☺ = _____

**MyMaths**.co.uk

Q 1205    SEARCH

# 8e Reading and drawing bar charts

1  Use the data to fill in this bar chart.

| Fruit | Apple | Banana | Pear | Orange |
|---|---|---|---|---|
| Frequency | 4 | 8 | 2 | 6 |

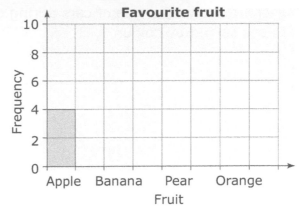

**Favourite fruit**

2  Use the data to complete this bar chart.

| Number of pets | 0 | 1 | 2 | 3 | 4 |
|---|---|---|---|---|---|
| Frequency | 14 | 10 | 4 | 6 | 2 |

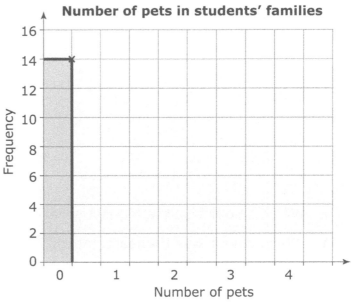

**Number of pets in students' families**

3  Use the data to draw a bar chart.

| Bird | Frequency |
|---|---|
| Jay | 5 |
| Robin | 5 |
| Magpie | 10 |
| Crow | 20 |
| Sparrow | 15 |

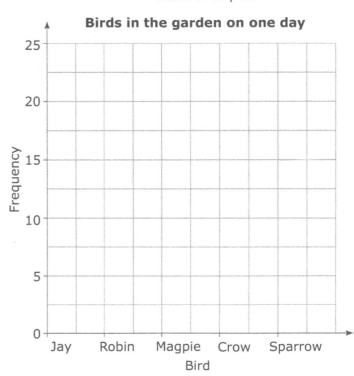

**Birds in the garden on one day**

## 8g  Reading diagrams

1   This bar chart shows the sale of cars during one week.
    The cars are sorted by colour.

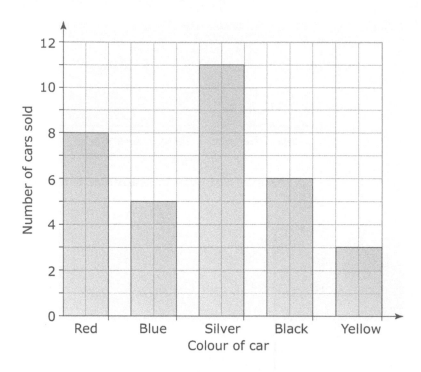

a   Which colour was the biggest seller? _____

b   Which colour was the least popular? _____

c   5 cars of one colour were sold. What colour were they? _____

d   How many cars were sold in total? _____

2   Draw the data from the bar chart onto this pictogram.
    Use the key to help you.

| | Number of cars sold |
|---|---|
| Red | |
| Blue | |
| Silver | |
| Black | |
| Yellow | |

Key:

= 2 cars

= 1 car

**MyMaths**.co.uk

1198, 1205, 1206   SEARCH

1   This bar chart shows the number of books sold in a book
    shop on Monday.

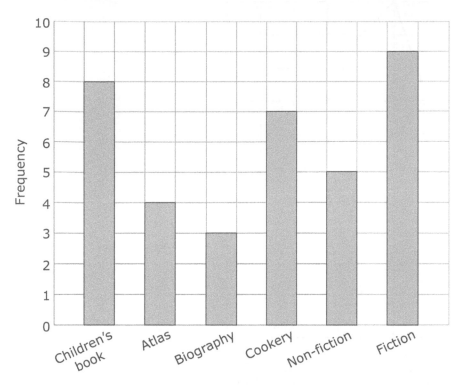

Types of book sold

**a** Put the books in order of sales starting with the **fewest** sold.
   **Biography, Atlas,** _____

**b** How many cookery books were sold? _____

**c** How many children's books were sold? _____

**d** How many atlases were sold? _____

**e** How many more children's books than atlases were sold? _____

**f** What was the **least** popular type of book sold? _____

**g** Which type of book was the **most** popular? _____

**h** The **mode** is the type of book that was most popular.

   The mode is _____

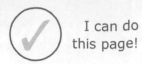
These are the average temperatures each month for London and Wick.

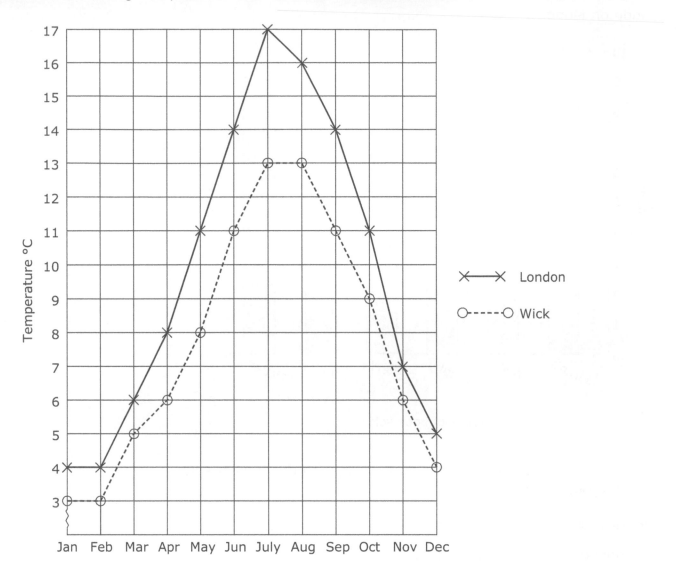

1 Does the line like this ×——× represent London or Wick?_____

2 What does the Wick line look like? _____

3 Why are there 12 points on the graph for each city? _____

4 What is the temperature range? _____

5 Which place had the highest temperature in July? _____

6 What was the highest July temperature? _____

7 What was the average Wick temperature in March? _____

8 What is the difference in temperatures in May? _____

9 What is the difference in temperatures in July? _____

MyMaths.co.uk

Q 1200, 1203 SEARCH

# 9a  Lines of symmetry

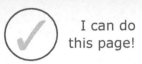
Draw all the lines of symmetry onto these shapes.

Remember to use a ruler.

6 mm

The lines of symmetry
are dashed lines.

**1**

**2**

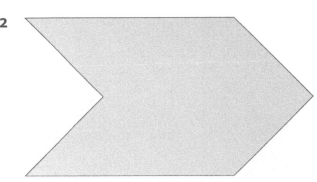

**3**

**4**

**5**

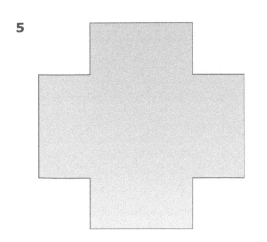

**6**

1 Draw the reflection of each shape in the mirror line.

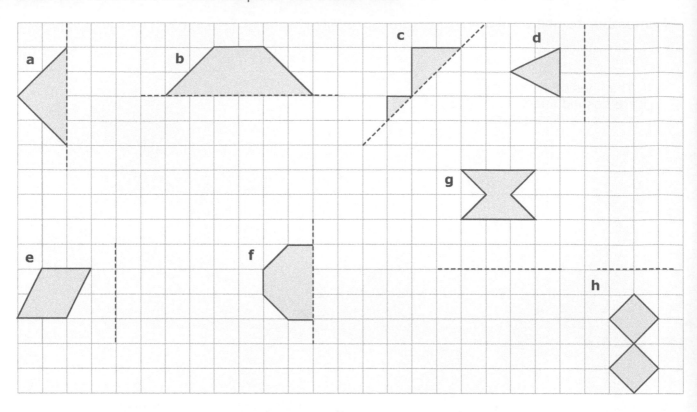

2 Reflect this drawing in the mirror line to complete the 'happy' face.

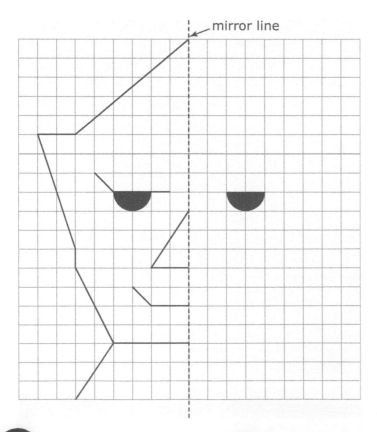

mirror line

MyMaths.co.uk

Q 1114, 1230 SEARCH

I can do this page!

**1** Use these instructions to translate each shape and draw its new position.

The first one is done for you.

Shape **a**

8 right and 3 up

Shape **b**

5 right and 2 down

Shape **c**

5 left and 2 down

Shape **d**

5 left and 3 up

Shape **e**

2 left and 4 up

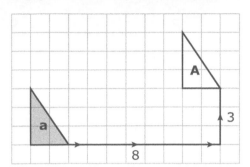

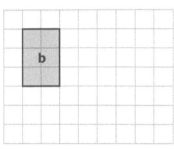

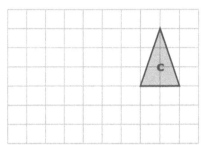

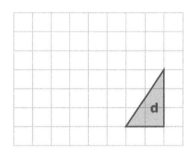

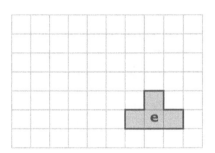

**2** Describe the translation of each shape.

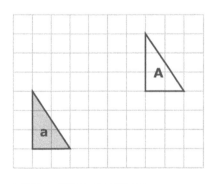

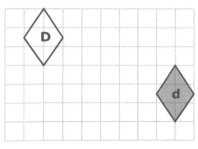

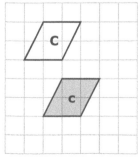

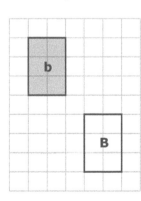

**a** to **A**    6 right and _____ up

**b** to **B**    3 right and _____ down

**c** to **C**    _____
                 _____

**d** to **D**    _____
                 _____

Count across first then up or down.

1  How many degrees are shown in these angles?

a

_____°

b

_____°

c

_____°

d

_____°

2  One wheel is turning clockwise and the other is turning anticlockwise.
   Label each wheel correctly.

_____

_____

3  Rotate this shape through 180°.
   Draw the new shape.

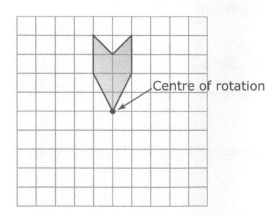

Centre of rotation

4  a  Rotate this shape **A** through 90° clockwise.
      Label the new shape **B**.

   b  Now, rotate the original shape **A** through
      90° anticlockwise.
      Label the new shape **C**.

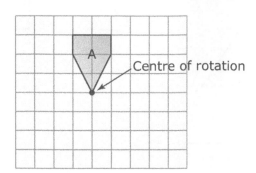

Centre of rotation

# 9e Tessellations

1   Tessellate this shape 15 times on the grid.

It must fit together with no gaps or overlaps.

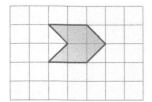

2   Which of these shapes will tessellate? Use the grid to test each shape.

If the shape tessellates, put a tick (✓) inside it.

a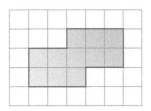

b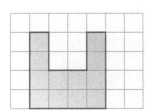

c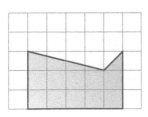

d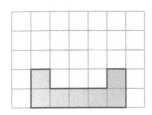

e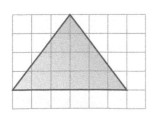

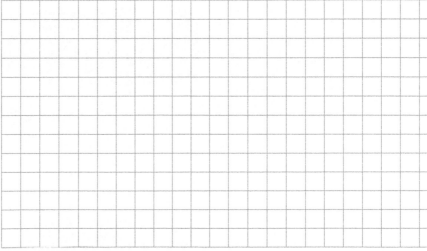

A Rangoli pattern is a Hindu design traditionally made during the Diwali festival to welcome guests. These simple instructions show you how to make a Rangoli pattern.

**Task 1**

**a** Draw a horizontal and a vertical axis on square dotty paper. Draw a simple design in the top left-hand quadrant.

**b** Reflect the lines in the horizontal axis and then reflect the whole pattern in the vertical axis.

**c** Draw diagonal lines through the origin. Turn your paper so these lines are axes and draw your pattern in the new top left-hand quadrant.
Repeat step 2.

**d** Erase the axes and diagonal lines. Colour in the regions if you want to.

**Task 2**
Draw the lines of reflection symmetry on your pattern.

**Task 3**
Make up your own design as in step 1 of Task 2. Create a Rangoli pattern from your design.

**1** Use the rule in the machine to complete each number mapping.

**a**

**b**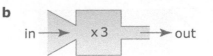

**c**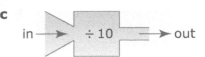

| in | → | out |
|---|---|---|
| 3 | → | 7 |
| 16 | → | ☐ |
| 32 | → | ☐ |

| in | → | out |
|---|---|---|
| 2 | → | 6 |
| 10 | → | ☐ |
| 30 | → | ☐ |

| in | → | out |
|---|---|---|
| 30 | → | ☐ |
| 50 | → | ☐ |
| 120 | → | ☐ |

**2** For each mapping:

▶ Use pairs of values to decide on the rule.

▶ Write the rule into the machine.

▶ Use the rule to complete the mapping.

**a**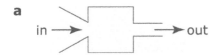

**b**

**c**

| in | → | out |
|---|---|---|
| 3 | → | 12 |
| 5 | → | ☐ |
| 10 | → | 40 |

| in | → | out |
|---|---|---|
| 15 | → | ☐ |
| 10 | → | 6 |
| 28 | → | 24 |

| in | → | out |
|---|---|---|
| 30 | → | 15 |
| 50 | → | ☐ |
| 12 | → | 6 |

**3** This is the rule that connects the number of white and grey beads in a necklace.

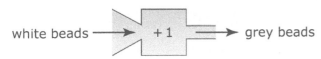

Shade the grey beads on the drawing using the rule.

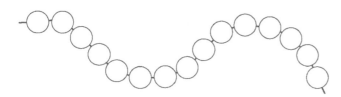

1  Fill in the operation that undoes each of these operations.

**a**  +3

**b**  x 4

**c**  − 7

**d**  ÷ 5

2

> F moves a robot forwards.
> F3 means go forwards 3 squares.
> R means turn right.
> L means turn left.
> T means turn around.

To move the robot from
A to B you write:

F3 − R − F2

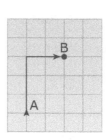

To take the robot back you
reverse all the instructions:
F2 − L − F3

Draw the path of the robot for:

**a**  F2 − R − F1 − R − F3

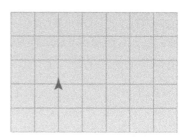

**b**  F1 − L − F3 − R − F2

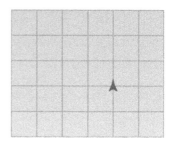

3  For each diagram:

▶  Give instructions to move the robot from A to B.

▶  Give the reverse instructions to move the robot back to B.

**a**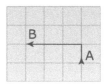

A to B: _____

B to A: _____

**b**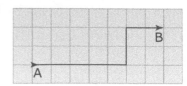

A to B: _____

B to A: _____

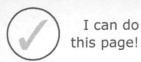
1  Calculate the weights of the parcels needed to make the scales balance.
   Write the weights onto the parcels.

**a**
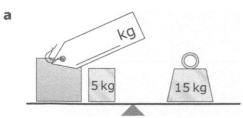

**b**
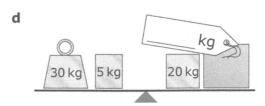

**c**
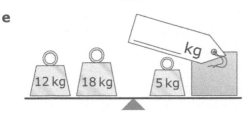

**d**
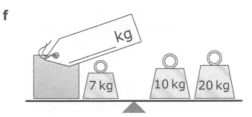

**e**
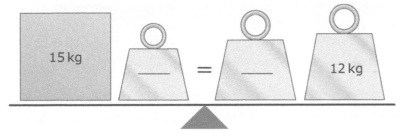

**f**

2  Write weights onto this drawing to make these scales balance
   (there is more than one correct answer).

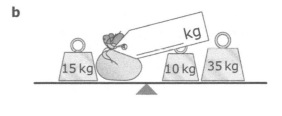

3  Write weights onto these parcels to make the scales balance.

**a**
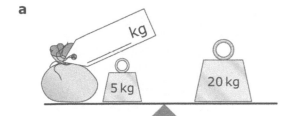

**b**

1  Work out the missing weights of the parcels on these scales and write them on.

a
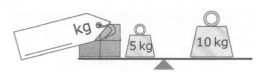

b
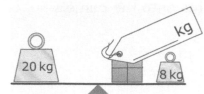

c
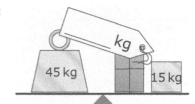

d
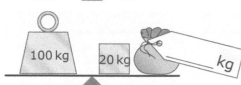

e
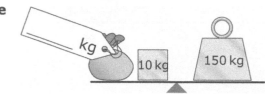

f
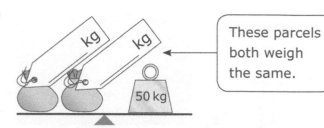

These parcels both weigh the same.

2  Use your answers to question 1 to find the value of each letter.
   The first one is done for you.

a  $a + 5 = 10$

   $a = 10 - 5$

   $a = 5$

b  $b + 8 = 20$

   $b = 20 - 8$

   $b = \underline{\quad}$

c  $c + 15 = 45$

   $c = 45 - 15$

   $c = \underline{\quad}$

d  $d + 20 = 100$

   $d = \underline{\quad} - \underline{\quad}$

   $d = \underline{\quad}$

e  $e + 10 = 150$

   $e = \underline{\quad} - \underline{\quad}$

   $e = \underline{\quad}$

f  $2f = 50$

   $f = 50 \div 2$

   $f = \underline{\quad}$

MyMaths.co.uk

1925  SEARCH

# 11a Factors

You can arrange these 10 counters into two rectangle patterns.

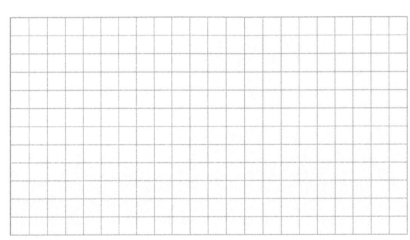

2 × 5          or          10 × 1

10 = 2 × 5 and 1 × 10

1, 2, 5 and 10 are **factors** of 10.

**1 a** Arrange these 12 counters into three different rectangle patterns on this grid:

**b** List all the factors of 12: _____

**2 a** There are 18 counters here.

Draw rectangle patterns to show all the factors of 18.

There are 3 different rectangle patterns.

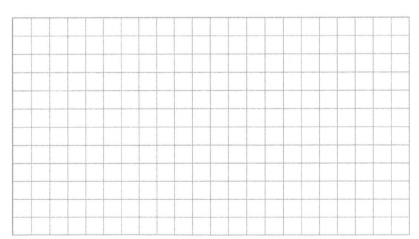

**b** List all the factors of 18: _____

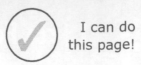

1 Multiply the numbers in the arrows by 3.
Write your answers in the circles.

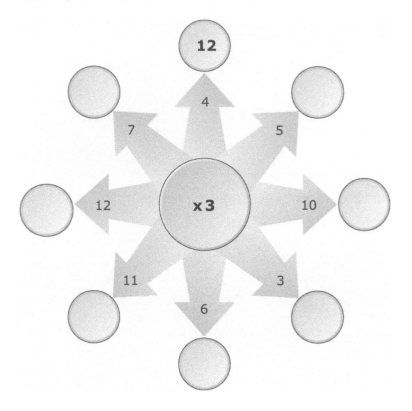

These are all multiples of 3.

2 Multiply the numbers in the arrows by 4.
Write your answers in the circles.

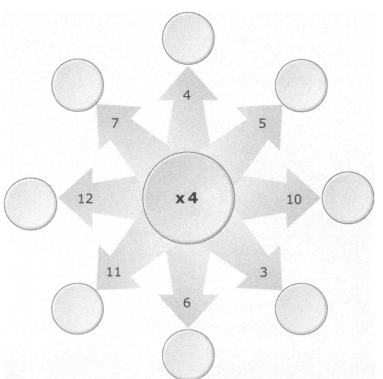

These are all multiples of 4.

MyMaths.co.uk

Q 1035 | SEARCH

# 12a  3D shapes

Fill in the table for these shapes

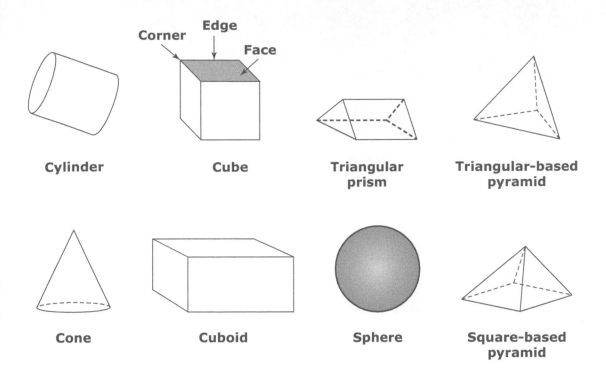

| Shape | Faces | Edges | Corners |
|---|---|---|---|
| cylinder | 3 | 2 | 0 |
| cube | | | |
| cone | | | |
| cuboid | | | |
| triangular prism | | | |
| triangular-based pyramid | | | |
| square-based pyramid | | | |
| sphere | | | |

What would you do if you got lost in a maze like this?

## Task 1

Keep your RIGHT hand on the wall and you are bound to get out!

Continue this line next to the right hand wall and see if you get through the maze.

## Task 2

I thought you were meant to use your LEFT hand?

Continue this line next to the left hand wall and see if you get through the maze.

## Task 3

a) Do the left hand and right hand methods both get you through this maze?

b) Can you find a shorter route through the maze?

## Task 4

Draw your own maze.

Try to include lots of dead ends and false trails.

Can your friend find a way through your maze?

## Task 5 (extension)

Find out about famous mazes such as the one at Hampton Court.

Can you find your way around them?

1 Here is a pattern that grows. Complete the last two parts of this pattern.

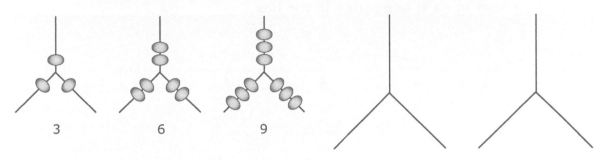

3        6        9        ____       ____

2 Each set of cards shows a number pattern.
Complete the pattern.

a  | 2 | 4 | 6 | 8 | __ | __ |

b  | 5 | 10 | 15 | 20 | __ | __ |

3 This drawing shows 1 table and 5 chairs:

a Complete the third drawing in this sequence:

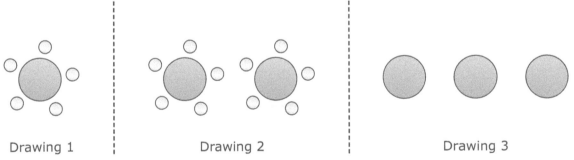

Drawing 1            Drawing 2                Drawing 3

b Complete this sentence: 'Each table has _____ chairs'.

4 a Add the next drawing in this sequence.

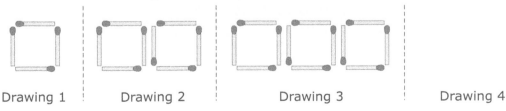

Drawing 1     Drawing 2          Drawing 3          Drawing 4

b Complete this sentence: 'Each square uses _____ matches'.

# 13b  Describing sequences

 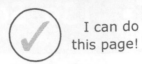

**1** Write the next number in each **sequence** in the box.

**a**

0  1  2  3  4  5  6  7  8  9  10  11

**b**

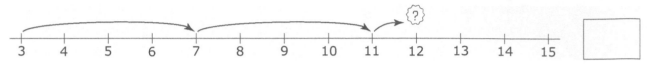

3  4  5  6  7  8  9  10  11  12  13  14  15

**c**

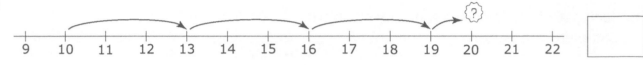

9  10  11  12  13  14  15  16  17  18  19  20  21  22

**2** Complete each sequence and describe the pattern.
Here are some sentences to help you.

> This sequence starts at _____.
>
> This sequence increases by _____ each time.
>
> This sequence decreases by _____ each time.

**a**  1, 4, 7, 10, _____, _____          This sequence _____

**b**  10, 20, 30, 40, _____, _____          This sequence _____

**c**  0, 4, 8, 12, _____, _____          This sequence _____

**d**  3, 8, 13, 18, _____, _____          This sequence _____

**e**  24, 21, 18, 15, _____, _____          This sequence _____

**3** The **rule** of a sequence tells you how to get to the next number.
Use the **rule** to write the first five numbers in each sequence.

**a**  Start at 0. The rule is +3.          _____, _____, _____, _____, _____

**b**  Start at 2. The rule is +2.          _____, _____, _____, _____, _____

**c**  Start at 7. The rule is +5.          _____, _____, _____, _____, _____

I can do this page!

1 Start at the bottom and follow the sequence to work out who is who!

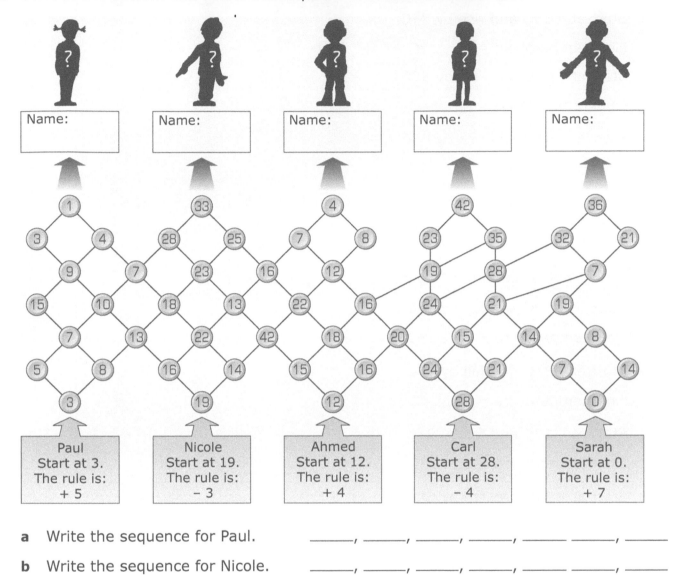

Name:

Name:

Name:

Name:

Name:

① 1
③ 3  ④ 4
⑨ 9  ⑦ 7
⑮ 15  ⑩ 10
⑦ 7  ⑬ 13
⑤ 5  ⑧ 8
③ 3

㉝ 33  ㉕ 25
㉘ 28  ㉓ 23  ⑯ 16
⑱ 18  ⑬ 13
㉒ 22  ㊷ 42
⑯ 16  ⑭ 14
⑲ 19

④ 4
⑦ 7  ⑧ 8
⑫ 12
㉒ 22  ⑯ 16
⑱ 18  ⑳ 20
⑮ 15  ⑯ 16
⑫ 12

㊷ 42
㉓ 23  ㉟ 35
⑲ 19  ㉘ 28
㉔ 24  ㉑ 21
⑮ 15  ⑭ 14
㉔ 24  ㉑ 21
㉘ 28

㊱ 36
㉜ 32  ㉑ 21
⑦ 7
⑲ 19
⑧ 8
⑦ 7  ⑭ 14
⓪ 0

| Paul | Nicole | Ahmed | Carl | Sarah |
|------|--------|-------|------|-------|
| Start at 3. | Start at 19. | Start at 12. | Start at 28. | Start at 0. |
| The rule is: | The rule is: | The rule is: | The rule is: | The rule is: |
| + 5 | − 3 | + 4 | − 4 | + 7 |

a   Write the sequence for Paul.      _____, _____, _____, _____, _____ _____, _____

b   Write the sequence for Nicole.    _____, _____, _____, _____, _____ _____, _____

c   Write the sequence for Ahmed.     _____, _____, _____, _____, _____ _____, _____

d   Write the sequence for Carl.      _____, _____, _____, _____, _____ _____, _____

e   Write the sequence for Sarah.     _____, _____, _____, _____, _____ _____, _____

2   Write the start number and rule for each sequence.

a   5, 7, 9, 11, 13, 15, …        Start at _____      The rule is _____.

b   2, 7, 12, 17, 22, 27, …       Start at _____      The rule is _____.

c   20, 18, 16, 14, 12, 10, …     Start at _____      The rule is _____.

d   20, 30, 40, 50, 60, 70, …     Start at _____      The rule is _____.

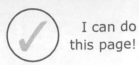
1 Find the final temperature. Use the thermometer to help you.

   **a** Start at −5°C and rise by 7°C

     Final temperature = _____°C

   **b** Start at −3°C and rise by 5°C

     Final temperature = _____°C

   **c** Start at −4°C and rise by 8°C

     Final temperature = _____°C

   **d** Start at −7°C and rise by 5°C

     Final temperature = _____°C

2 Final the final temperature. Use the thermometer to help you.

   **a** Start at −5°C and fall by 3°C

     Final temperature = _____°C

   **b** Start at −3°C and fall by 4°C

     Final temperature = _____°C

   **c** Start at −1°C and fall by 8°C

     Final temperature = _____°C

   **d** Start at −2°C and fall by 6°C

     Final temperature = _____°C

3 Fill in the missing numbers from these sequences.

   **a** Start at −8 and add 3.

     −8, −5, _____, 1, _____, _____, 10, _____

   **b** Start at 5 and subtract 2.

     5, _____, _____, −1, _____, −5, _____

 MyMaths.co.uk    Q 1173    SEARCH

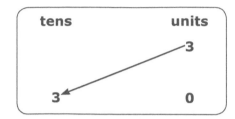
When you multiply a number by 10 you move the digits one place to the left:

$3 \times 10 = 30$          $16 \times 10 = 160$

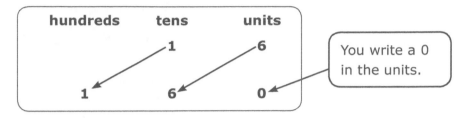

| tens | units |
|------|-------|
|      | 3 |
| 3 | 0 |

| hundreds | tens | units |
|----------|------|-------|
|          | 1 | 6 |
| 1 | 6 | 0 |

You write a 0 in the units.

1  Multiply these numbers by 10. Put the digits in the correct place.
Remember to put a 0 in the empty units place.

|   |          | hundreds | tens | units |
|---|----------|----------|------|-------|
| a | 6 × 10 = |          |      |       |
| b | 9 × 10 = |          |      |       |
| c | 5 × 10 = |          |      |       |
| d | 8 × 10 = |          |      |       |
| e | 10 × 10 = |         |      |       |

|   |           | hundreds | tens | units |
|---|-----------|----------|------|-------|
| f | 12 × 10 = |          |      |       |
| g | 23 × 10 = |          |      |       |
| h | 40 × 10 = |          |      |       |
| i | 51 × 10 = |          |      |       |
| j | 83 × 10 = |          |      |       |

2  Double each of the numbers in the arrows.

Write your answers in the circles.

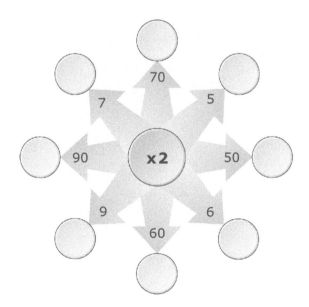

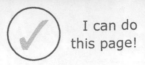

**1** Complete these calculations.

    **a** 2 × 4 = _____     **b** 3 × 6 = _____     **c** 5 × 6 = _____

    **d** 8 × 7 = _____     **e** 5 × 4 = _____     **f** 6 × 8 = _____

    **g** _____ × 4 = 16     **h** 3 × _____ = 24     **i** _____ × 4 = 36

**2** Split up these numbers into tens and units.

    **a** 16 is the same as ☐10☐ + ☐☐     **b** 11 is the same as ☐10☐ + ☐☐

    **c** 19 is the same as ☐10☐ + ☐☐     **d** 13 is the same as ☐10☐ + ☐☐

**3** Fill in the boxes to multiply these numbers.

    **a** 16 × 4    16 is ☐10☐ + ☐☐     **b** 19 × 3    19 is ☐☐ + ☐9☐

             10 × 4 = ☐☐                  10 × 3 = ☐☐

              6 × 4 = ☐☐                   9 × 3 = ☐☐

       So     16 × 4 = ☐40☐ + ☐☐      So     19 × 3 = ☐☐ + ☐☐

                     = ☐☐                        = ☐☐

    **c** 15 × 7    15 is ☐10☐ + ☐☐     **d** 14 × 4    14 is ☐☐ + ☐☐

             10 × 7 = ☐☐                  10 × 4 = ☐☐

              5 × 7 = ☐☐                   4 × 4 = ☐☐

       So     15 × 7 = ☐☐ + ☐☐      So     14 × 4 = ☐☐ + ☐☐

                     = ☐☐                        = ☐☐

**MyMaths**.co.uk

Q 1024, 1904   SEARCH

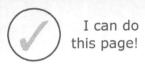

I can do this page!

When you divide a number by 10, you move the digits one place to the right.

$\frac{320}{10}$ = 32    hundreds    tens    units

3          2          0

3          2

Divide these numbers by 10. Move the digits one place to the right.

The first one is done for you.

| | calculation | hundreds | tens | units |
|---|---|---|---|---|
| **1** | | | 9 | 0 |
| | 90 ÷ 10 = | | | 9 |
| **2** | | | 3 | 0 |
| | 30 ÷ 10 = | | | |
| **3** | | | | |
| | 50 ÷ 10 = | | | |
| **4** | | | | |
| | 120 ÷ 10 = | | | |
| **5** | | | | |
| | 170 ÷ 10 = | | | |
| **6** | | | | |
| | 230 ÷ 10 = | | | |
| **7** | | | | |
| | 540 ÷ 10 = | | | |
| **8** | | | | |
| | 860 ÷ 10 = | | | |
| **9** | | | | |
| | 490 ÷ 10 = | | | |
| **10** | | | | |
| | 800 ÷ 10 = | | | |

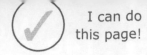

Use each number line to work out the division.

The hints are there to help you.

**1**  32 ÷ 4 = ☐

−4

| | | | | | | | | |
|0|4|8|12|16|20|24|28|32|

Jump backwards until you reach 0.

**2**  40 ÷ 5 = ☐

| | | | | | | | |
|0|5|10|15|20|25|30|35|40|

**3**  27 ÷ 3 = ☐

| | | | | | | | | | | |
|0|3|6|9|12|15|18|21|24|27|30|33|

Be careful where you start.

**4**  25 ÷ 5 = ☐

25

Write your own numbers on the line.

Count down in 5s.

**5**  48 ÷ 4 = ☐

Write your own numbers on the line.

Start at 48 and count down in 4s.

**MyMaths**.co.uk

🔍 1022, 1391  SEARCH

# 14g  Written methods of division

 I can do this page!

Fill in the spaces to help you complete the divisions.

**1**  93 ÷ 3

```
      9  3
  –   3  0        ←——— 3 × 10
      6  3
  – [      ]      ←——— 3 × 10
    [      ]
  – [      ]      ←——— 3 × 10
        [    ]
  –     [    ]    ←——— 3 × 1
```

93 ÷ 3 = [      ]

**2**  192 ÷ 6

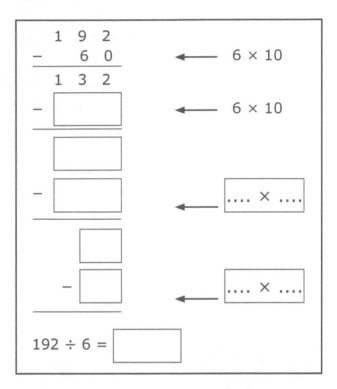

```
      1  9  2
  –      6  0        ←——— 6 × 10
      1  3  2
  – [        ]       ←——— 6 × 10
    [        ]
  – [        ]       ←——— [ .... × .... ]
        [     ]
  –     [     ]      ←——— [ .... × .... ]
```

192 ÷ 6 = [      ]

**3**  108 ÷ 9

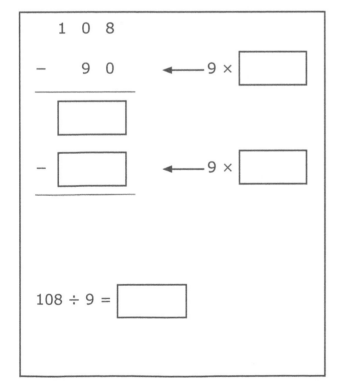

```
      1  0  8
  –      9  0      ←—— 9 × [    ]
    [        ]
  – [        ]     ←—— 9 × [    ]
```

108 ÷ 9 = [      ]

**4**  92 ÷ 4

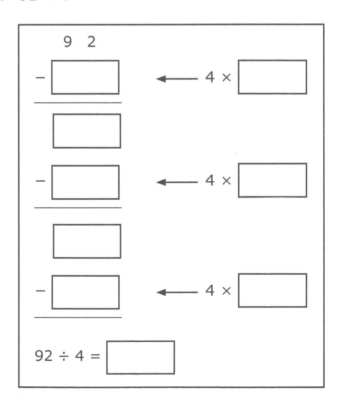

```
      9  2
  – [      ]      ←—— 4 × [    ]
    [      ]
  – [      ]      ←—— 4 × [    ]
    [      ]
  – [      ]      ←—— 4 × [    ]
```

92 ÷ 4 = [      ]

**MyMaths**.co.uk

Q 1021, 1041  **SEARCH**

61

# Case Study 5: Electricity in the home

Have you ever thought about how many things around the house use electricity?
As energy costs rise, more people are keeping an eye on their electricity usage.

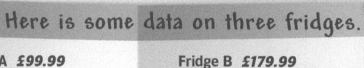

## Here is some data on three fridges.

**Fridge A** *£99.99*

Fresh food storage
❄ volume 86 litres

Freezer compartment
❄ volume 10 litres

Energy efficiency
❄ class 'A'

Energy consumption
❄ 139 kWh per year

**Fridge B** *£179.99*

Fresh food storage:
❄ volume 245 litres

Freezer compartment
❄ none

Energy efficiency
❄ class 'A'

Energy consumption
❄ 164 kWh per year

**Fridge C** *£299.99*

Fresh food storage
❄ volume 122 litres

Freezer compartment
❄ volume 18 litres

Energy efficiency
❄ class 'A'

Energy consumption
❄ 234 kWh per year

**Electricity costs 15p per kWh**

?

**Task 1**

a Look at the energy consumption figures for each fridge. Work out the annual cost of running each fridge. Give your answer in pounds and pence.

b Which fridge has the most space for fresh food?

c Which fridge would you choose and why?

**Task 2**

The average annual electricity usage per household in the UK is 3300 kWh (kilowatt hours).

a If electricity costs 15p per kWh, how much would a typical household pay for their electricity per year?

b How much would this work out as per month?

62

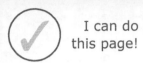

**1 a** Colour 4 of these tins red and 3 tins blue.

**b** What is the ratio of red to blue? _____ : _____

**2 a** Colour 3 of these tins red and 6 tins blue.

**b** What is the ratio of red to blue? _____ : _____

**3** What is the ratio of blue beads to white beads on these strings?

**a** _____ : _____

**b** _____ : _____

**c** _____ : _____

**d** _____ : _____

**4** This string of beads is made up from blue and white beads in the ratio of 1 : 4.
Colour in the string of beads.

**5** In this string the ratio of blue beads to white beads is 3 : 2.
Colour in this string of beads.

**6** The ratio of blue beads to white beads on a string is 2 : 1.
There are 10 blue beads.
How many white beads are needed? _____

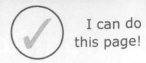

The perimeter of a shape is the total distance around the edge.

This triangle has sides of 4 cm.

Its perimeter is 4 cm + 4 cm + 4 cm = 12 cm

You can say: $3 \times 4\,\text{cm} = 12\,\text{cm}$

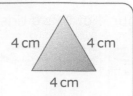

1 Find the perimeter of these shapes.

**a**

The perimeter is:

_____ cm + _____ cm + _____ cm

= _____ cm

Or:

_____ × _____ cm = _____ cm

**b**

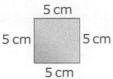

The perimeter is:

_____ cm + _____ cm + _____ cm + _____ cm

= _____ cm

Or:

_____ × _____ cm = _____ cm

2 In these questions the lengths of each side are written in symbols.

You can add them just like numbers. Find the perimeter of each shape.

**a**

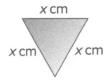

The perimeter is:

_____

= _____ cm

Or:

_____ × _____ cm = _____ cm

**b**

The perimeter is:

_____

= _____ cm

Or:

_____ × _____ cm = _____ cm

**MyMaths**.co.uk

🔍 1037, 1393 **SEARCH**

# 16a Introducing probability

1   Read the sentences and decide if the events are certain, uncertain
    or impossible.

    Put the correct letter after the sentence.

    C certain                    U uncertain                    I impossible

| | | |
|---|---|---|
| a | All cars will need new tyres. | |
| b | All cars will need new tyres tomorrow. | |
| c | Cars have number plates. | |
| d | Cars have 4 doors. | |
| e | All buses are red. | |
| f | Buses carry passengers. | |
| g | You sit on a bus. | |
| h | Trains travel along the road. | |
| i | Trains have 4 carriages. | |
| j | Planes fly in the air. | |
| k | Trains travel on rails. | |
| l | Roads have yellow lines. | |
| m | Cars travel at the speed limit. | |
| n | Boats are used for fishing. | |
| o | Children enjoy playing with toy cars. | |
| p | A space ship will land in the park. | |

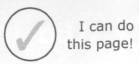 I can do this page!

When you spin two coins, there are four ways they can land.

 or  or  or

Follow these steps to see if you can guess which way will come up!
You will need two coins.

**1** Guess a result – two heads, two tails or a head and a tail. You can write: HH, TT or HT.

**2** Spin two coins.

**3** If the guess is correct follow the tick (✓) to the next level. If not, follow the cross (✗).

**4** Repeat until you reach the bottom level. This shows your percentage of correct guesses.

You can play the game with other students – take it in turns to guess and use a different colour to mark your results.

START HERE ➡

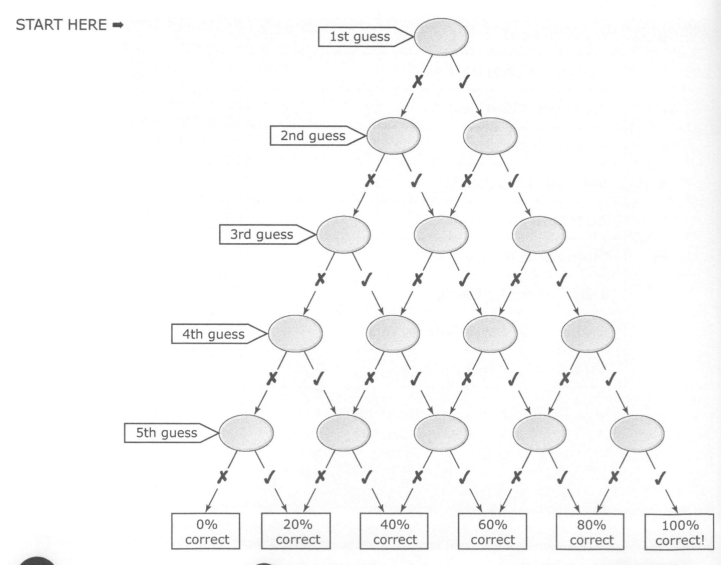

MyMaths.co.uk     🔍 1209     SEARCH

1  If I choose card **5**
   I have to pay ................
   I have a ..... in ...... chance of winning.
   I have a ..... in ...... chance of losing.

2  If I choose card **2**
   I have to pay ................
   I have a ..... in ...... chance of winning.
   I have a ..... in ...... chance of losing.

**SPIN TO WIN!**
To play the game, pay 50 p and choose a card.
The spinner will spin when all 5 numbers have been bought.
The person with the number the spinner stops at wins a fabulous prize!

**a prize every game!**

**50p per go!**

If the stallholder gave money as a prize, how much do you think it should be?

3  If I choose cards **1** and **3**
   I have to pay ................
   I have a ..... in ...... chance of winning.
   I have a ..... in ...... chance of losing.

4  To be certain of winning, I would have to choose cards
   ..........................................................
   I would have to pay ........................

# Checklist – I can do it!

| Page | | Title | I can do it! |
|------|------|------|------|

# Multiplication table

| ×  | 1  | 2  | 3  | 4  | 5  | 6  | 7  | 8  | 9  | 10  |
|----|----|----|----|----|----|----|----|----|----|-----|
| 1  | 1  | 2  | 3  | 4  | 5  | 6  | 7  | 8  | 9  | 10  |
| 2  | 2  | 4  | 6  | 8  | 10 | 12 | 14 | 16 | 18 | 20  |
| 3  | 3  | 6  | 9  | 12 | 15 | 18 | 21 | 24 | 27 | 30  |
| 4  | 4  | 8  | 12 | 16 | 20 | 24 | 28 | 32 | 36 | 40  |
| 5  | 5  | 10 | 15 | 20 | 25 | 30 | 35 | 40 | 45 | 50  |
| 6  | 6  | 12 | 18 | 24 | 30 | 36 | 42 | 48 | 54 | 60  |
| 7  | 7  | 14 | 21 | 28 | 35 | 42 | 49 | 56 | 63 | 70  |
| 8  | 8  | 16 | 24 | 32 | 40 | 48 | 56 | 64 | 72 | 80  |
| 9  | 9  | 18 | 27 | 36 | 45 | 54 | 63 | 72 | 81 | 90  |
| 10 | 10 | 20 | 30 | 40 | 50 | 60 | 70 | 80 | 90 | 100 |